QUALITY ASSURANCE FOR
THE CHEMICAL AND PROCESS INDUSTRIES
A Manual of Good Practices
Second Edition

Also available from ASQ Quality Press

Specifications for the Chemical and Process Industries: A Manual for Development and Use
ASQ Chemical and Process Industries Division,
Chemical Interest Committee

ISO 9000 Guidelines for the Chemical and Process Industries, Second Edition
ASQ Chemical and Process Industries Division

Guidelines for Laboratory Quality Auditing
Donald C. Singer and Ronald P. Upton

ANSI/ISO/ASQC Q9000-1994 Series: Quality Management and Quality Assurance Standards

To request a complimentary catalog of ASQ publications, call 800-248-1946.

QUALITY ASSURANCE FOR THE CHEMICAL AND PROCESS INDUSTRIES
A Manual of Good Practices
Second Edition

American Society for Quality
Chemical and Process Industries Division
Chemical Interest Committee

ASQ Quality Press
Milwaukee, Wisconsin

Quality Assurance for the Chemical and Process Industries, Second Edition

Library of Congress Cataloging-In-Publication Data

Quality assurance for chemical and process industry: a
    manual of good practices / prepared by American
    Society for Quality, Chemical and Process Industries
    Division, Chemical Interest Committee—2nd ed.
      p. cm.
    Includes bibliographical references and index.
    ISBN 0-87389-439-1 (pbk.: alk. paper)
    1. Chemical industry—Quality control. I. American
    Society for Quality. Chemical Interest Committee.
    TP149.Q34   1998
    660'.068'5—dc21   98-35423
                                       CIP

10 9 8 7 6 5 4 3 2 1

ISBN 0-87389-439-1

Acquisitions Editor: Ken Zielske
Project Editor: Annemeike Koudstaal

ASQ Mission: The American Society for Quality advances individual and organizational performance excellence worldwide by providing opportunities for learning, quality improvement, and knowledge exchange.

Attention: Schools and Corporations
ASQ Quality Press books, audiotapes, videotapes, and software are available at quantity discounts with bulk purchases for business, educational, or instructional use. For information, please contact ASQ Quality Press at 800-248-1946, or write to ASQ Quality Press, P.O. Box 3005, Milwaukee, WI 53201-3005.

For a free copy of the ASQ Quality Press Publications Catalog, including ASQ membership information, call 800-248-1946.

Printed in the United States of America

 Printed on acid-free paper

American Society for Quality

ASQ

Quality Press
611 East Wisconsin Avenue
Milwaukee, Wisconsin 53201-3005
800-248-1946
Web site http://www.asq.org

Contents

Preface

The Chemical and Process Industries Division of the American Society for Quality (ASQ) is committed to quality improvement throughout the Chemical and Process Industries. Since 1984, the Chemical Interest Committee (CIC) has worked to reach consensus on and to document good quality practices for our industries. In 1987, through the auspices of the Quality Press, the Committee published the first edition of this book *Quality Assurance for the Chemical and Process Industries—A Manual of Good Practices*. Affectionately referred to as the "Little Red Book," it has been a best seller for the Quality Press.

When the ISO 9000 standard for quality systems was first released in 1987, the CIC saw a need to link the concepts of good practices in its publication with the new standard. In 1992, the Committee published an ISO guidelines book. This has also been a best seller for the Quality Press. A second edition of *ISO Guidelines for the Chemical and Process Industries* was released in 1996. The CIC also prepared for Quality Press an additional volume, *Specifications for the Chemical and Process Industries—A Manual for Development and Use*, that was released in 1996.

This second edition has been updated to be consistent with the 1994 ANSI/ISO/ASQC Q9000 revisions and other developments in quality practice over the decade since its original publication. The revision committee has reviewed, revised, updated, and made minor additions. The guiding theme for the revision followed a frequently heard comment, "I like the small manual. It's all there. Don't change the character if you revise it." The bibliography has been updated and expanded. An index has been added.

Acknowledgments

ASQ and the C&PID Chemical Interest Committee appreciate the contributions of the authors of this revision. They are

Bradford S. Brown	Consultant
Georgia Kay Carter	Hercules Incorporated
Rudy Kittlitz (chair)	DuPont

We would also like to acknowledge the major contribution of those who wrote the original version of the book and offered suggestions for this revision:

David Anderson
Gary Anderson
Jim Bigelow
Brad Brown
Georgia Kay Carter
Ken Chatto
Louis Coscia
David Files
Peter Fortini
Richard Hoff
Sherman Hubbard
Rudy Kittlitz
Norman Knowlden
Anil Parikh
David Stump
Jack Weiler

Dedication

Richard C. Hoff, 1926–1998

The authors wish to dedicate this revised edition of *Quality Assurance for the Chemical and Process Industries* to the memory of Richard C. Hoff. Dick was a Senior Member of ASQ and a Certified Quality Auditor. He was a member of the Chemical Interest Committee from the beginning and served as chair for the preparation of the first edition of *ISO Guidelines for the Chemical and Process Industries*. He was a good friend to all in the committee and a role model of quality process. He will be missed.

1. Purpose

This manual documents good quality practices in the chemical and process industries. It is intended for quality professionals, process engineers, manufacturing managers, suppliers, and customers of the chemical and process industries to use as a reference to build, guide, or evaluate the suitability of a quality system.

Perspective:

The art and science of sampling and measurement are different in chemical and process industries as compared to mechanical industries. The relatively large contribution of measurement variability requires proper use of statistical methods for process control. Because of this, proper interchange and understanding of information between customers and suppliers become very important.

Additional considerations requiring special attention and a different framework to apply quality management methods in the chemical and process industries are listed as follows

- Raw materials are often natural materials whose consistency depends on natural forces. This requires compensating adjustments in the process to maintain a consistent quality level of the finished product.
- The technology of chemical sampling presents complex problems because of chemical considerations and the varied nature of product units, ranging in weight from a few grams to thousands of tons.
- Chemical processes occur on the molecular level and changes are often inferred from secondary information.
- Measurement itself is a complex process that must be carefully standardized and controlled.
- Automatic process control is highly developed with computerized feedback and feed-forward loops. Emphasis is on "prevention."
- Once produced, product properties often change with time and with changes in environmental conditions.
- There may not be a direct relationship between the measured properties and the performance characteristics of the product in use.
- True defectives are rarely found, although nonuniformity between lots is often detected by the customer.
- Safety, health, and environmental impact of both the process and the product are major factors in quality management.

We have tried to focus on these issues and to balance them with the needs of the industry and of its customers for sound quality practices. This is not a "stand alone" document, and a bibliography has been included.

2. Organization and Responsibilities

Summary:

- An organization should document its quality policy, quality system, and quality planning requirements in a manual or series of manuals. The Quality Manual should clearly state policy, define procedure, and specify responsibility for quality system components.

- An organization should periodically measure and report progress toward meeting quality objectives.

- Constant innovation and superior customer service create excellence. Organizations must learn to develop flexibility while maintaining consistency. Innovation and quality improvement are different processes from quality control, and they result in change rather than maintenance of status quo.

- Safety, health, and environmental issues impact both process and product in chemical and process industries. An organization should document these policies and procedures in a manual or a series of manuals that specifies responsibility for these system components.

2.1 Quality Policy

The following has been taken from "Quality Management and Quality System Elements—Guidelines," ANSI/ASQC 9004-1-1994, Sections 4.1 and 4.2:

> "The responsibility for and commitment to a quality policy belongs to the highest level of management. Quality management encompasses all activities of the overall management function that determine the quality policy, objectives, and responsibilities, and implement them by means such as quality planning, quality control, quality assurance, and quality improvement within the quality system.

> "The management of an organization should define and document its quality policy. This policy should be consistent with other policies within the organization. Management should take all necessary measures to ensure that its quality policy is understood, implemented, and reviewed at all levels of the organization."

2.2 Quality Systems

The sharing of product data with the customer should not take place unless there is a functioning quality management system in place.

Management of quality systems involves

- Management responsibility
- Statistical quality control
- Knowledge of measurement variability, process variability, and "tools'" to keep variability under control

- Documented procedures for consistent treatment of information, that is, data
- A system for continual improvement

2.3 Quality Manual

The quality manual should include the following

2.3.1 A description of how quality will be designed into the manufacturing process and product, including

- Raw material control systems
  - —Vendor selection and certification procedures
  - —Lot sampling and acceptance practices
  - —Test methods
- Process control systems and practices
- Testing, analysis, and product acceptance systems
  - —Good laboratory practices
  - —Documented test methods
  - —Sampling plans
  - —Sampling procedures
  - —System for maintaining reference materials
  - —Standardization/calibration
  - —Lot acceptance and product control systems
- Product shipment systems
  - —Packaging and labeling quality assurance
  - —Shelf life and inventory management
  - —Shipment container quality assurance
- Nonconforming material disposition
  - —Recall/retrieval practices
  - —Nonconforming raw material
  - —Nonconforming product

2.3.2 A definition of customer relations practices.

- Policy for sharing information
- Method for analyzing and resolving complaints, including corrective and preventive action procedures
- Method for determining customers' needs
- Corrective action for process upsets and nonconforming product

4

2.3.3 A description of regulations that impact company operations.

2.3.4 Documentation of procedures for developing and commercializing new products.

2.3.5 A description of standard operating procedure and its administration.

2.3.6 A definition of record-retention practices.

2.3.7 Designation of responsibilities and contacts for quality functions.

2.3.8 Specifications.
- Purchase
- Process
- Product
- Package and label
- Sales

2.3.9 A description of the quality audit system and practices.

2.4 Quality Reporting

2.4.1 Process quality data used to make process control decisions should be available to operating personnel, on a real-time basis. Summary reports of process performance should be available for line management to provide awareness, identify improvement opportunities, and allocate resources.

2.4.2 Reports should be made to management showing progress against company quality objectives. Areas for reporting include customer perception of quality and value, controlling processes to target, and process capabilities. The purpose of management reports is to inform and to assist in the allocation of resources. Quantitative measures should be used whenever possible.

2.5 Quality Improvement

2.5.1 Quality improvement.
- Assessment
  - —Existing quality performance level and variability of process (used in its broadest sense) should be measured statistically over time.
- Process control and correction
  - —Remove error and inconsistency (special causes).

—Reduce variability. Emphasize on-line measurement and on-line control.

—Maintain closeness to target.

- Process improvement

—Must be led by management and must involve all employees.

—Requires formal periodic review, assessment, and organization to develop that improvement.

—Occurs when process knowledge is increased. Procedures such as experimental design, screening studies, evolutionary operations, analysis of variance, analysis of means, and the like, should be used to increase process understanding.

—Occurs when teams, trained in problem solving and statistical tools, cut across departmental barriers and work together.

2.6 Customer Service and Product Performance

- Superior customer service requires a marketing system that can determine and translate customer needs and desires into actual product performance and customer service.

- Process data collection and analysis should ultimately be integrated into systems linking it to the product's field performance.

2.7 Safety, Health, and Environment

The safety, health, and environmental policies should be documented.

- Operational guidelines to handle processes and products safely and responsibly

- Safety rules for manufacturing and nonmanufacturing areas

- A description of safety, health, and environmental audit system and practices

- A description of regulatory matters that affect company operations

- An approach to provide pertinent safety information to company product users, for example, material safety data sheets (MSDS).

3. Basis for Specifications

Specifications are the agreed upon, documented requirements between customer and supplier. They should include sampling and testing methods, limits, targets, and reporting requirements. For a fuller discussion of specifications, see *Specifications for the Chemical and Process Industries*, also prepared by the C&PID Chemical Interest Committee.

Summary:

- Guidelines for determining product specifications should be provided. Whenever possible, specifications should be in terms of true value, that is, the required property value if measured without measurement variability.

- The purpose of obtaining a sample from a production lot is to provide information for an effective decision regarding acceptance or rejection using the lot acceptance plan. The sample should be representative of the entire lot. It can be either a random sample or a systematic sample. In general, it is preferable to have a minimum sample size of two, with the samples taken at distinctly different locations in the lot. This will allow estimates of the within-lot variability to be used in designing lot acceptance plans. Caution must be used to ensure that the samples are independent and represent within-lot variance and not just measurement variability.

- Good quality practice should be documented concerning interpretation of numerical data and specification tolerances, which are the essential elements for agreement on the critical measurements for product acceptance. These are based on the general procedure and conventions of Food Chemical Codex, United States Pharmacopoeia, National Formulary, American Society for Testing and Materials, and American Chemical Society (Reagent Chemicals).

3.1 Definition of Terms

3.1.1 *Specification.* A document that contains a precise statement, or set of requirements (quantities, characteristics, properties, or service) to be satisfied by a material, process, or product with notations, as appropriate, of the procedure(s) that should be used to determine if the requirements are satisfied. As far as practicable, each requirement should be expressed numerically as a target value with required limits.

3.1.2 *Target.* Agreed upon value within the specification limits that produces an optimal result for the customer and/or supplier. The target need not be the center of a specification range.

7

3.1.3 *True value.* A quantitative characteristic that does not contain any sampling or measurement variability.

3.1.4 *Product variability.* Amount of variability in a set of measured values caused by actual differences in product characteristics from product unit to product unit or from time to time. From the customer's viewpoint, the supplier's product variability manifests itself as variability in the performance of the customer's process as the customer consumes the material.

3.1.5 *Purchase specification.* A formal document stating all of the information necessary for the purchase of a process material, including

- Target value and limits for all characteristics that define the required product quality
- The specific test method for each characteristic
- Packaging and shipping requirements
- Safety, health, and environmental requirements
- Applicable federal and state regulations
- Terms for rejection of material

3.1.6 *Sales specification.* A formal document containing the information necessary to describe a product for a particular customer use and end use.

3.1.7 *Lot.* A definite quantity of a product or material accumulated under conditions considered uniform for sampling purposes. For a continuous process, a lot is usually defined by a time interval. For a batch process, each batch can be considered a lot. This definition is based on the broad definition of a process as any activity that changes the quality parameters used to define the product: blending, mixing, packaging under different conditions, and so forth, as well as the usual production processes.

3.1.8 *Process capability.* The variability of a product property when the process is in a state of statistical process control.

3.1.9 *Process performance.* The variability of a product property as produced over an extended period of time. Process performance consists of process capability plus all other sources of variability experienced during the time period.

3.2 Determining Product Specifications

3.2.1 Product specifications, listed in preferred order.

- Whenever possible, specifications should be stated in terms of true product variability—that is, the required property values if measured without measurement variability.

- Customer requirements, preferably determined by technically sound procedures, should be the primary basis for specifications.

- If customer requirements have not been determined, or if customer and supplier targets differ, specifications must be developed and agreed to by the supplier and customer.

- The supplier's demonstrated process performance should be the starting point for specifications. The customer's experience with this product should also be considered. If the customer's experience has been satisfactory, specifications should be set close to the supplier's demonstrated performance.

- If the supplier's process has not been producing satisfactory product, specifications should be agreed to that the supplier can meet and the customer is willing to accept. The supplier and customer should also agree on goals and timing for supplying satisfactory product. This may involve process improvements or the use of a lot acceptance plan for product release.

- Temporary specifications should be set for new products or when there are insufficient supplier or customer data available to determine specifications. When sufficient data become available, these specifications should be reviewed for any required changes.

3.2.2 Changing specifications.

- Specifications based entirely on customer requirements should be changed only when customer requirements change.

- Negotiated specifications should be changed as follows

  —Specifications based on process performance may be reviewed for change when process performance improves or when customer requirements have been determined.

  —Specifications determined from insufficient data should be reviewed for change when sufficient data become available.

- Specifications should be reviewed for change when suppliers or customers make any change in their processes that could affect the adequacy of current specifications.

3.3 Incoming Material Acceptance and Product Release

3.3.1 Incoming material acceptance.

Incoming material acceptance procedures should be developed, taking into consideration process requirements and supplier capability. The goal is to minimize or eliminate the need for formal incoming lot acceptance and to rely on the supplier to supply products that continually meet specifications. Supporting documentation may be required in the form of certification, data sharing, or statistical evidence. Minimal inspection or testing may still be required for identification or to detect changes occurring during shipment.

Formal lot acceptance procedures may be required based on prior experience with the material or because there is a need for risk avoidance.

3.3.2 Lot acceptance for product release.

If the existing process control is not sufficient to ensure shipping products that continually meet specifications, it may be necessary to use a lot acceptance procedure.

3.4   Sampling

Samples taken from incoming materials acceptance may be for lot against specifications or only for identification. Samples taken from in-process materials or from final product may be for process control or for lot acceptance against specifications or both. When samples are used for both process control and product release the sample frequency is determined by the most critical application. When the sampling is done for the purpose of lot acceptance, each lot should be sampled by taking a representative sample either randomly or systematically.

In general, it is preferable to have a minimum sample size of two, taken from distinctly different locations within the lot; this will allow ongoing estimates of the within-lot variability used in designing effective lot acceptance plans.

3.4.1 Discrete units.

When the lot is made up of discrete units such as bags, drums, bales, or tubes of yarn, the sample should consist of material from at least two different units. A random sample is often adequate for incoming lots of material. However, a systematic sample is preferred when the units are in some form of a time sequence, the same as for product being released for shipment. The systematic sample should be distributed throughout the lot so that the within-lot variability is included.

3.4.2 Bulk product.

When the lot is a single unit of bulk material such as high- or low-viscosity liquids, powders, pellets, or particles, the sample should be taken from at least two locations in the lot. The sample locations can be chosen randomly, and the individual samples may be grab samples, thief composite samples, or stratified samples—that is, top, middle, and bottom.

3.4.3   Blends.

Blends apply primarily to product lots being released for shipment. Blends are used to combine material from two or more in-process sources for efficient utilization of equipment.

- If blending is used to reduce the effect of in-process differences, good quality practice dictates modifying the process to eliminate these differences so that blending is not required.

- If specifications are on the blended product, the sample should be taken on the blended material.

3.5   Measurement Variability and Acceptance Decisions

3.5.1   Acceptance decisions are made based on observed or measured value. The observed value is the sum of a true value and a random variation from the measurement process. The total variance of observed values is equal to the variance of the true values, plus the variance of the measurement errors. In the chemical and process industries measurement variability is often a large proportion of the total observed variability. It is not uncommon for the measurement variance to be 50 percent of the total observed variance.

3.5.2   Test methods that are accepted industry-wide are preferred whenever possible. In any case, when electing the test method to be used, the effect of measurement variability on the reported result should be considered. Measurement variability can be viewed as consisting of repeatability and reproducibility.

- Repeatability refers to the variability in the reported results measured at the same time, on the same equipment, by the same operator, and so forth.

- Reproducibility refers to the additional variability in the reported results measured at different times, on different equipment, by different operators, and so forth.

3.5.3   Replicate testing can reduce the effect of measurement variability on the final reported result. If one replication is done within a short period of time, such as one after another, only the part of the measurement variability associated with that time period is reduced. However, if a large source of measurement variability is from day-to-day, the effect of this variability can only be reduced by replication over days. This may indicate an out-of-control measurement system. A time series, component of variance analysis may be useful in planning the replication schedule.

3.5.4 The amount of measurement variability affects decisions made from measurement results. This is particularly important in the chemical and process industries. Limits used to decide if product meets specifications should be determined using a procedure based on the operating characteristic curve (OC curve) of the test. The shipping and/or acceptance limits can then be selected, balancing the risks of accepting materials that do not meet specifications and rejecting the materials that meet specifications. Replicate measurements can be used to change the shape of the OC curve and thus change the risk.

3.5.5 When the test method includes a precision statement, customers using well-calibrated equipment should accept the product unless their acceptance values differ from the supplier's reported value by more than expected, based on the precision statement.

3.6 Acceptance Plan Design

- Acceptance plans for either incoming materials or product release use the same design criteria. The acceptance decision is made on a lot-by-lot basis. For incoming materials, an entire shipment may be a single lot, or the shipment may be divided into one or more lots for acceptance purposes. For products to be shipped, the lot may be a single unit or a predetermined number of units.

- Acceptance plan design (sample size and limits) must take into account the effect of within-lot and measurement variability. This is of utmost importance when the measurement variability is a large portion of the total observed variability. This will involve making use of the operating characteristic (OC) curve for each plan.

  Standard acceptance plans may be found in the following references:

  —ANSI/ASQC Z1.4–1993. Sampling Procedures and Tables for Inspection by Attributes.

  —ANSI/ASQC Z1.9–1993. Sampling Procedures and Tables by Variables for Percent Nonconforming.

  These standard plans may not be readily adaptable to many chemical operations.

3.7 Operating Characteristic Curve

The OC curve is a plot for a specified acceptance sampling plan that shows the probability of accepting a sampled lot for any given true level of the parameter being measured. The parameter being measured may be either an attribute or a variable.

The OC curve is used to select acceptance limits and a sample size that allows for within-lot and measurement variability with predetermined risks of accepting material outside of specifications and rejecting material that meets specifications. See chapters 12 and 13 of *Specifications for the Chemical and Process Industries,* for a fuller discussion of OC curves.

3.8 Risks of Error

- Alpha risk is the probability of making a Type I error, that is, rejecting a lot that is conforming or within specifications. This is sometimes referred to as the producer's risk.

- Beta risk is the probability of making a Type II error, that is, accepting a lot that is nonconforming or outside of specifications. This is sometimes referred to as the consumer's risk.

3.9 Record Retention

3.9.1 Retention time.

All data related to the lot of any product produced for sale or internal use should be maintained for whichever of the following is longer.

- The stated useful life of the specific lot
- One year after the final sale of the specific lot
- Regulatory requirements (e.g., FDA)
- Customer requirements (e.g., QS-9000)

All information not related to a specific product should be maintained for at least two years.

3.10 Number Rounding

3.10.1 Rounding of numerical data.

To round a number to a given number of significant figures, consider the digit in the next place beyond the one to be retained. Then follow rounding procedures as described in ASTM Standard E 29, Section 3. In a string of calculations, do not round until the last step.

3.10.2 Significant figures for tolerance limits.

- When limits are expressed numerically, the upper and lower limits of a range are inclusive so that the range consists of the two values themselves and all the intermediate values, but no values outside the limits.

- For specifications, values are considered significant to the last digit shown. For example, a requirement of not less than 96.0 percent would be met by a result of 95.96 percent but not by a result of 95.94 percent. Similarly, a specified range of 46 percent to 49 percent means that only values between 45.5 and 49.4 will comply (following rounding procedures as in Section 3.10.1).

13

3.10.3 Examples of rounding off.

Acceptance Limit	Observed Value	If Rounded Off to Nearest	Correct Rounded Off Value	Complies with Acceptance Limit
	59,940	100 psi	59,900	No
60,000 psi, min.	59,950	100 psi	60,000	Yes
	59,960	100 psi	60,000	Yes
	56.4	1%	56	No
57%, min.	56.5	1%	56	No
	56.6	1%	57	Yes
	0.54	0.1%	0.5	Yes
0.5%, max.	0.55	0.1%	0.6	No
	0.56	0.1%	0.6	No

3.10.4 Unit conversion in measurement systems.

- Accurate conversions are obtained by multiplying the specific quantity by the appropriate conversion factor given in the ASTM Standard E 29 or the National Bureau of Standards (NBS) Handbook 44, 1979.

- In terms of significant figures, the rule for multiplication and division is that the product or quotient should contain no more significant digits than are contained to the right of decimal in the number with the fewest significant digits used in the multiplication or division.

- Example of conversion:

  $113.2 \times 1.43 = 161.876$, which rounds to 161.9.

4. Sampling Technology

Summary:

- Sampling is the process of removing a portion of a material for analysis. Ideally, the sample should truly represent the material from which it was taken. It must maintain its integrity with respect to the characteristics measured, and the act of sampling should not change the material sampled. Sampling should provide effective measurement information about the material with known and controlled risks.

- Sampling technology considers the sources of sampling variation with respect to making effective decisions.

- The physical state of the material affects the sampling process.

- The environment, including containers, of the material to be sampled must be considered in order to protect samplers from the material and to protect the material from contamination, change, or degradation during sampling.

- Samples may be taken for lot acceptance or release, for process control, and for many other reasons. The sampling approach may differ, depending on whether the characteristics measured are attributes or variables. Sampling choices include 100 percent inspection, sampling plans, or no sampling.

- Training is recommended to develop personnel with the skills and knowledge needed to take samples.

- Retaining samples is generally a good practice, but except for certain regulatory requirements, sample retention is not an absolute requirement. There is a balance between retaining all samples for a long time to cover all eventualities and the real economic problem of storage space, especially when a special environment is required because of product instability.

4.1 Sampling Variation

4.1.1 Definition of population.

In the chemical and process industries the population sampled for product acceptance is usually a lot (see Section 3.1.7).

For process control, the population sampled is generally a function of process time.

4.1.2 The effect of sample size on decisions.

- Decisions regarding process control or product acceptance depend upon the homogeneity of the sample for a characteristic measured for a lot.

- The level of the characteristic for the lot is estimated as the average of the sampling measurements of that characteristic taken on that lot. The samplings must be sufficient to represent the variability of that characteristic throughout the lot.

- The homogeneity of the lot is estimated from the variability found between measurements on the representative samplings taken and from the sampling variation.

- Increasing the number of samplings taken and measured will provide greater confidence in estimating the level and in estimating the sampling variance. (An increase in sampling does not decrease the sampling variance.)

- Increasing the number of measurements taken on each sampling can be used to reduce the contribution of measurement variability to the overall sampling variance.

4.1.3 Components of Sampling variation.

- Between samplings

  The variability detected between measurements of a characteristic on replicate samplings of a lot provides a measure of the lot homogeneity, within-sample variability, and measurement variability.

- Within a sampling

  Replicate determinations of a characteristic on the same sampling provide a measure of the within-sample variability and measurement variability.

4.1.4 Homogeneity considerations.

- Process knowledge

  —Knowledge of the materials and process involved should be used to select the sampling procedure that minimizes sampling costs and variability.

  —Supporting information on lot definition, process capability, run order, supplier certification, process control systems, and quality audit results should also be used.

- Discrete units

  The representative sample should consist of material obtained from discrete units distributed throughout the lot so that within-lot variability is included in the sample.

- Bulk material

  The representative sample should consist of material obtained from locations distributed throughout the lot so that within-lot variability is included in the sample.

Within lot variability tends to increase proceeding from single ingredient gases through low-viscosity liquids and solutions; single-ingredient uniform size solids to high-viscosity liquids and emulsions; to multiple particle size solids. The number of lot locations sampled should increase accordingly.

4.1.5 Composite samples.

Compositing of multiple physical samples for analysis is commonly used to reduce testing cost. The compositing of physical samples is appropriate if

- The measurement variability is small
- The within-lot variability is large
- Only an estimate of the average is required

Compositing is NOT good practice. It is appropriate only if all three conditions are met. Better practice calls for individual analyses on each sample. If the additional testing cost or volume is not acceptable, then alternative sample structures or test methods should be investigated.

A major problem with compositing is that it conceals the within-lot variability. The absence of an estimate of within-lot variability may mask and defer needed corrective action to reduce the within-lot variability.

4.2 Sampling Method

There should be a written description of the sampling process and a statement of the use to be made of the results for every sampling operation. The sampling procedure should contain:

- The safety, health, and environmental warning labels for the material to be sampled.
- The environment requirements to maintain the integrity of the material and its samples.
- A listing of sampling equipment, tools, and container requirements.
- A written procedure on how the sample is to be obtained, in order to ensure consistency.
- A definition of the exact location of each sampling point.
- The sample size needed for testing, number, and weight or volume. Note that multiple samples are not always economically or physically feasible.
- A list of the analyses to be performed.

4.3 Physical State and Source

4.3.1 The physical state describes the state of the material as a liquid, gas, solid, or mixture of these states. The source of the material to be

17

sampled can be in pipes, conveyors, piles on the ground, storage bins, tanks, rail cars, tank wagons, drums, bottles, bales, and the like. Both the physical state and the source affect the sampling equipment used, the equipment needed to protect the material from the environment, the homogeneity of the material, and the number of samples required.

4.3.2 In a system containing particulates, it is desirable to take samples when the system is in motion.

4.3.3 Sampling systems should be designed to avoid "dead legs"; that is, fresh, current material must be dispensed by the sample system, not material that has been sitting in the system.

4.4 Sampling Conditions

4.4.1 The proper sampling conditions include arrangements for safety, health, and other considerations that impact on the quality of the material in either its bulk or sample form. Examples: toxic materials, materials absorbed through skin, materials under high pressure, and materials sensitive to air, light, moisture, heat, or cold.

- The technician must be knowledgeable of potential hazards associated with each material sampled and use appropriate protective equipment.
- The sampled material must be protected from exposure to contamination.

4.4.2 Container.

- Material of composition

   The container and its closures must neither react with nor be soluble in the sample. Examples: Glass is generally inert, but soft glass can contaminate a sample with sodium; plastics have additives that can affect samples; solvents can permeate polyethylene; caps, covers, and cap liners can also be sources of contamination.
- Cleanliness

   Cleanliness appropriate to the analysis is mandatory. New containers may contain fibers, dust, and dirt. Washed containers may contain detergent residues. Tap water can contaminate certain samples, and piping or tubing for distilled or deionized water can also be a contamination source. Appropriate conditioning and/or purging of sample lines and containers may be required before taking the sample.

4.4.3 Equipment.

Sampling equipment is an ever-important consideration. Construction, design, and suitability of material must be carefully considered.

4.5 Technician Training

The technician who takes the samples must be trained and proficient in the specific sampling procedure, equipment, and tools used. It is helpful if the technician has knowledge of the chemistry of the materials being sampled and has basic understanding of the statistical principles underlying sampling procedures.

4.5.1 Sampling tools.

The reasons for tool design and the method of use need to be explained. Use of the tools should be a part of the training of technicians. Demonstration of the ability to use the tools should be a requirement to pass the training.

4.6 Sample Management

4.6.1 Labeling and records.

All samples should be labeled so that they are traceable to their bulk source, to the technician, and to the date and time they were taken. Labels should meet regulatory requirements. Records should be accurately kept and should be identifiable with each individual sample.

4.6.2 Incoming samples.

Incoming samples should be stored so as to preserve their integrity before analysis. It is never safe to assume that a sample received by the laboratory is homogenous.

4.7 Retained Samples

4.7.1 Samples may be retained for the following reasons

- For compliance with the law.
- To resolve differences between vendor and customer. Whenever possible, material from the actual containers involved should be tested, and a retained sample used only as a backup or second choice sample.
- For studies concerning changes that have occurred in a process that cannot be related to any known factor.
- To establish a data base for a new or modified test.
- For periodic review of product stability, product or process drift, and the like.
- To establish representative values for properties not normally measured.

4.7.2 Quantity of retained samples.

An amount at least twice as large as the amount necessary to run all the routine lot acceptance requirements is desirable.

4.7.3 Retention time.

The minimum retention time should be at least the shelf life of the lot or one year after the final sale of the lot. Special environments may be required for long-term storage.

4.7.4 Storage Containers.

The container for retained samples should be appropriate for the retention time and product characteristics under study. It is desirable for the container to represent the normal product package as closely as possible and to have no adverse effect on the material.

4.7.5 Security and storage conditions.

A suitable system should be used to control access to retained samples to protect their validity and to preserve the original quality.

4.7.6 Hazardous and reactive products.

Further requirements and procedures may be needed for the retention of hazardous or reactive products.

4.7.7 Regulatory requirements.

Any regulatory requirement that specifies retained samples shall be followed and be considered minimal good practice for any material manufactured under these requirements.

5. Analysis and Testing

Summary:

Any laboratory producing data for process control or final lot acceptance should meet the following requirements as a minimum for operation. The operation of the laboratory should be recognized as a process, and controlled and managed as such.

- Good quality practice requires documentation of test methods and general preference for international and/or national standard methods over nonstandard methods.

- Good quality practice for standards, controls, and traceability should be maintained. Calibration standards should be used to set test equipment values before actual testing. Controls are used to monitor the performance of test methods.

- Certification of material can be expressed in several forms. The most common are Certificate of Analysis (Test) or Certificate of Conformance. The titles "Quality Control Report" or "Quality Assurance Report" are sometimes used to cover the same information. The documents are used to provide the customer with information regarding a given shipment.

- To be of value, a certificate must accompany the shipment or be sent under separate cover, arriving before or soon after receipt of the product shipment. The timing should be negotiated with the customer.

5.1   Good Laboratory Practices

5.1.1   Personnel.

- Records should be retained covering
    —Job descriptions or other definitive declaration of individual authority and responsibility for all personnel
    —Internal training of all related individuals
- Types of individuals required to operate a laboratory:
    —*Technician.* Individual who carries out analyses and has demonstrated an ability to read, understand, and perform the prescribed procedures with documented training.
    —*Technical resource.* Individual with expertise and advanced training related to the prescribed procedure.
    —*Laboratory supervisor.* Individual responsible for coordinating daily activities, training, documentation, and the general running of the laboratory within company guidelines. This function may be incorporated in a team model.

5.1.2   Documentation of operations.

Records should be retained covering

- Equipment maintenance
- Calibration and standardization including appropriate statistical control charts
- Corrective and preventive action activities related to out-of-control conditions
- Laboratory data with traceability to the technician who produced the results
- Test methods and laboratory procedures, both current and historical, with approval for change, revision dates, and so forth
- Sample schedules and methods

5.1.3   Interlaboratory activities.

When two or more laboratories are testing or measuring the same characteristic on a given material, it is recommended that regular, controlled interlaboratory comparisons be conducted. The participating laboratories may be the supplier, customer, independent laboratories or regulatory agencies.

5.2   Test Methods Documentation

Analytical and test methods used for purposes of process control or to determine conformity to material specification requirements are central to the quality system of a process industry. Each organization must establish procedures for the preparation, approval, and preservation of test method documentation. Exact procedures used should be well documented, including sampling, sample preparation, conditioning, testing, recording, and preparation for next test (cleanup).

5.2.1   Compendial methods.

Compendia of test methods have been developed and compiled by several standards organizations such as the International Organization for Standardization (ISO), American Society for Testing and Materials (ASTM), Association of Official Analytical Chemists (AOAC), United States Pharmacopoeial Convention, and others. Review processes and interlaboratory comparisons conducted by these organizations help to ensure the accuracy and reproducibility of the methods. Use of these methods, when available, is therefore recommended.

5.2.2   Nonstandard methods.

Deviation from the procedures specified by a compendial method are occasionally required because of limitations of available equipment, new equipment, or the properties of particular samples. When such non standard testing is to be conducted on a regular basis, or when legal or regulatory requirements are at issue, modified compendial methods should be documented as new methods.

5.2.3   Documentation of new methods.

Test methods other than established compendial methods must be documented as controlled documents. Essential elements of a test method include

- Date of approval of this revision
- Description of principle and validation of method
- Range of applicability of test results using this method
- All apparatus required for the test
- Reagents required by material name and grade or specification, Chemical Abstract Service (CAS) registry number, or by supplier and catalog number
- Instructions for preparation of standard calibrants, controls, solutions, and the like
- Safety consideration and hazards of materials and equipment used in the test
- Method of sample collection (see Section 4), protection, preparation, and conditioning
- Amount of sample to be used, together with a stated tolerance pertaining to the amount (e.g., 5.0 ± 0.1 g)
- Procedure for calibration of instruments
- Procedure for controlling the test, including sample control charts
- Step-by-step instructions for carrying out the test procedure and calculations, including replication requirements, if any
- Approximate time required for analysis
- Test interferences if known
- Precision statement on test repeatability and reproducibility, preferably in terms of standard deviations (as defined in ASTM D-3040)
- Procedure for making calculations and preparing the test report

5.3   Calibration and Traceability of Standards

5.3.1   Calibration.

Calibration is the statistically valid establishment that a measurement process reports values with the required uncertainty inside the range of

23

use and that the equipment detects changes in values with the uncertainty required for the decisions being made with the measurement results.

Calibration must be done in a statistically valid manner that establishes that the average measured value on the equipment being calibrated is equal to the reported measured value of the standard calibration material. Also, the variability of the equipment being calibrated must be determined so that the user of measured values is properly informed of the meaning of measurements made with the equipment.

Calibration may be repeated on a regular schedule, or calibration may be initiated only when a control chart signals a bias in the level of the test values.

5.3.2 Traceability.

- Each testing organization (normally a laboratory) should designate appropriate primary standards (calibrants) for particular materials, weights, and so forth, required for its work. The organization is responsible for protecting calibrants against loss, contamination, or change.

- It is desirable that standard materials be traceable to recognized (e.g., National Institute of Standard and Technology, NIST) sources. The essential requirement of good practice is consistency within and between the laboratories. If no recognized source is available, a mutually agreed secondary source may be used.

- When standards are used within a laboratory (for example, reagent solutions in ASTM E200), labeling for the standard should include at least

  —Identity of the material

  —Reference to the laboratory notebook recording details of its preparation

  —Expiration date beyond which the material should be discarded

5.3.3 Standards versus internal reference materials.

- In some cases, primary reference standards have been developed by governmental agencies such as the NIST. It is good quality practice to use primary reference standards traceable to third-party organizations or agencies when available. By definition, a primary reference standard has a unique given value, generally determined by an absolute method.

- More often, agreed upon internally generated reference materials or controls are used. These are often reagent grade or a carefully selected large sample from a well-characterized and mutually agreed upon lot or batch.

24

- These internal reference materials are used in many instances for interlaboratory (round robin) testing between customer and supplier. In general, they will have an agreed value and a confidence range. Internal reference materials (controls) are not calibrants.

- If a hierarchy of standards can be developed, that is, the measured values of the standard can be transferred mathematically from a primary material to a secondary working reference material, then this practice is acceptable to preserve what is usually a costly material.

- To measure drift with time and environment, internal reference materials are used to ensure that the test method is still on target, by retesting at specified intervals. Control charts should be used. The test method should provide guidelines to be followed.

5.4   Certification

5.4.1   Certificate of Analysis (COA).

A Certificate of Analysis indicates that testing was done on a lot or shipment to a customer. The actual contents of the certificate should be negotiated between the supplier and customer. The following elements are suggested as part of the shipment identification:

- Supplier order/reference number
- Customer order number
- Test results
- Date of shipment
- Product identity
- Product quantity
- Product origin
- Name and title of the certifying individual

For the test results:

- The actual test results for the specified lot should be given whenever possible. Supplying identification of the test method and the specification range for the tests is optional.

- For various reasons, the test may be on other than the lot specified. It may be on a statistical or physical composite of lots, may be a periodic sample, and so forth. When this is the case, the test result may be reported on the certificate, but the source of the test value should be identified.

5.4.2   Certificate of Conformance.

A Certificate of Conformance is an assertion that the product conforms with the given specification limits. The certificate relates to the contents of a given shipment (see Section 5.4.1), but provides no test results.

5.4.3 Certificate of Compliance.

A Certificate of Compliance is not the same as a Certificate of Conformance. A Certificate of Compliance is an assertion that the supplier of a product has met the requirements of a relevant specification, contract, or regulation. It makes no assertion concerning any given shipment.

We discourage the abbreviation COC since it is not clear whether it refers to Certificate of Conformance or Certificate of Compliance.

5.5 Independent Laboratories and Calibration Services

5.5.1 Purpose.

An independent laboratory is any testing analytical or calibration facility that is not under the direct management control of the company using the services of such independent laboratory. Independent laboratories are used when particular tests or methods cannot be performed conveniently, technically, economically, or in a timely manner by the company needing the data or test results.

5.5.2 Accuracy and precision responsibility.

It is the responsibility of the requesting organization to ensure that an independent laboratory has the capability, opportunity, and understanding necessary to meet the specific accuracy and precision requirements of the requested testing.

Before final commitment and at periodic intervals, the requesting organization should verify that the results of the independent laboratory meet established accuracy and precision criteria.

- The requesting organization should provide appropriate reference materials for internal use at the independent laboratory. If reference sample materials are subject to change with time or for other reasons, the requesting organization should detail the necessary storage conditions and/or provide timely replacement reference materials. The requesting organization should state the required precision on reference materials.

- The independent laboratory should periodically report the results of its statistical control data on reference materials to the requesting organization. The results should be reported after at least 50 tests have been run on the reference materials.

- Formal interlaboratory comparisons are appropriate means for auditing the quality partnership.

5.5.3 Qualification audit.

Before reaching any final agreement, the organization should audit the independent laboratory to verify that

- Adequate physical facilities and equipment exist to accomplish the test procedures
- Testing staff are trained and capable
- Sample retention facilities are adequate
- Adequate statistical control procedures for test methods and calibration are in place
- Proprietary information safeguards are adequate
- Management fully understands the agreed requirements

5.5.4 Agreement on services to be performed.

There should be a written agreement that fully describes the entire expectations of the receiving company and that documents the agreement of the independent laboratory to meet those expectations.

- The written agreement should include, but not necessarily be limited to, the following
  —Specific test methods to be used (ASTM or other)
  —Sample preparation responsibility
  —Sample identification details
  —Calibration requirements
  —Compliance with quality standards
  —Calculation of test results
  —Frequency of testing
  —Frequency of testing measurement control samples
  —Use of control charts
  —Repeat test requirements
  —Treatment of outliers
  —Location at which test will be performed
  —Details of sample transmittal
  —Details of result transmittal
  —Sample retention requirements
  —Protection of proprietary information

—Record keeping

—Notification of significant changes

• There should be a detailed, written procedure covering retest, resample, and recalibration procedures.

5.5.5 The independent laboratory should be provided with the name and phone number (with backups) of a qualified contact in the requesting organization who can handle technical or procedural questions.

6. Process Control

Summary:

- The quality of the equipment used in a chemical process has a major impact on the safety of the process and the quality of the product. The equipment must be capable of carrying out the process as planned, safely.

- Technological changes (including procedural, test method, formulation, and specification changes) may be made to a process for reasons of improved product quality, improved safety, or improved process efficiency. Product characteristics may be altered as a result of technological changes. Technological changes should be carefully controlled and documented to ensure that the product remains fit for the intended customer use.

- Quality is managed and assured in custom or research product manufacturing processes by controlling the raw materials, tools, and procedures of the process, together with a reliance on training of people.

- The use of control charts is important for statistical process control in the chemical and process industries.

- Process capability is defined, and guidelines are provided for use of the concept.

6.1 Process Documentation

6.1.1 Written instructions are used to describe in full the planned operation of any process. Quantitative terms should be used to describe all process parameters. Qualitative terms such as "hot–cold" and "acidic–basic" are not sufficient to control a material process. Qualitative terms do not facilitate the application of statistics for process development, control, and improvement—essential elements of good-quality written instructions.

Pre-startup considerations include

- Safety checklist and procedures by either explicit words or references to operating standards
- List of essential materials to be used, including qualities, quantities, and inspections
- Equipment inspection procedures

Process operation considerations include

- Step-by-step detailed description of the process:
  —Sequence of additions and process operations

—Points in the process where the material may be placed on hold without affecting the quality of the product

—Parameters of the process in quantitative terms with target values and limits

—Sampling schedule and procedures

—In-process testing requirements and the actions to be taken based on the results

—Container inspection and preparation procedures

—Process records (log sheets, notebooks) for

    —Raw materials—quantity used and lot number

    —Sequence and duration of additions

    —Parameters—temperature, pressure, stirring rates, flow rate, and duration

    —Samples taken—when, where, and how

    —Test results and actions taken

    —All other process actions and observations not specifically listed

- Abnormal operation plan

Instructions for proceeding in case of unplanned delays, emergency shutdowns, and other upsets should be documented.

6.1.2 Responsibility.

The production department is responsible for the preparation of process instructions. The process instructions should be written by a team made up of production, product development, product engineering, safety, analytical, and quality personnel. Each should contribute his or her respective expertise so that complete instructions will be provided to the production personnel.

6.2 Maintenance

6.2.1 Preventive maintenance programs are needed to maintain the quality and capability of the equipment. A preventive maintenance program includes

- Scheduled shutdowns for equipment inspections to detect weaknesses before they cause an unplanned process interruption, especially for those areas in which inspections cannot be performed without a shutdown

- Detailed procedures for shutdowns and checklists for inspections to ensure that vital steps are performed and vital points are checked every time

6.2.2 Repairs need to be made as soon as possible. Preventive maintenance promotes process and product uniformity/consistency.

- Installed spares may make it possible to effect repairs without a shutdown.
- Parts need to be available to make repairs in a timely manner to minimize the duration of scheduled, and especially unscheduled, shutdowns.
- Skilled personnel are needed to perform the inspections and repairs.

6.3 Technology Change Control

6.3.1 Responsibilities.

- Since the manager at the site has ultimate responsibility for product quality, he or she has the responsibility for control of current manufacturing practice and for technological changes.
- Functions such as product management, business management, quality assurance, engineering, development, and regulatory specialists should have input to technology change proposals.

6.3.2 Technology changes.

The following constitute technology changes that should be controlled

- Formulation change
  —Ingredients
  —Amounts
  —Sequence
- Process change in manufacturing and measuring
  —Equipment
  —Procedure
  —Conditions
- Testing change
  —Test conditions change
  —Test addition
  —Test deletion
  —Test frequency change
- Specification range change
- Shelf life change
- Packaging component change
- Label revisions
- New raw material qualification

- Site qualifications
  —Addition
  —Deletion
- Sampling plan change

6.3.3 Customer notification.

In general, customers should be notified of technology changes because of the potential for unforeseen product quality or performance effects. The decision for customer notification should be made jointly by marketing and manufacturing functions.

6.3.4 Documentation.

Agreed technology changes should be documented to provide traceability.

6.4 Custom or Research Production Control

6.4.1 Introduction.

The chemical and process industries span a spectrum of production from continuous high-volume products made routinely to specialty products produced in small quantities no more than once or twice a year. For specialty products, the application of good quality practices requires some specific consideration.

Often the production process is completed before sufficient samples can be tested and the results submitted. By the time the product is made again, the personnel and/or the equipment may have changed, and any process statistics would need to be redone. Thus, the concepts and applications of traditional process control and statistics are limited.

6.4.2 Process tools and conditions control.

Process tools and instrumentation require frequent standardization and statistical monitoring to demonstrate control (e.g., analytical instruments, scales, temperature controllers, flow meters, etc.). Their accuracy and precision must be determined. The goal is to reduce variation wherever possible.

Preventive maintenance programs are required.

6.4.3 Procedures control.

Process procedures are continuously subject to change. To minimize their impact on product quality, each must be rigorously evaluated (see Section 6.3). Changes for safety or changes in the analytical laboratory are process changes requiring similar evaluation.

6.4.4 Training.

Consistent, continual, current, and documented training is critical to control the variation contributed by personnel changes. Personnel must display the skills and personal characteristics that support safety and quality.

6.4.5 Supplier approval.

Long-term relationships with a few reliable suppliers are desirable. Sometimes, time does not permit the step-by-step supplier approval process needed for a new supplier. When this is the case, tightened inspection of incoming material may be required.

6.4.6 Process capability.

Process capability may be determined based on processes of a similar nature—for example, all materials produced by a general type of synthesis.

6.4.7 Process development.

Process optimization should be part of the development program. Experimental designs should be employed to improve the process continuously from run to run.

6.4.8 Sample evaluation.

Agreement between customer and supplier is heavily dependent on the evaluation of proposed product samples. The communication between the customer and supplier must be well established to determine which tests are needed to measure acceptance of product.

6.5 Use of Control Charts

6.5.1 Overview.

- Shewhart control charts (X, X-bar, R, s, MR) are often the best choice for process control if the required sampling is possible. They provide a widely understood framework of procedures to validate statistical process control.
- The Cumulative Sum (CUSUM) and the Exponentially Weighted Moving Average (EWMA) chart are frequently used for applications in the CPI where small shifts need to be detected as promptly as possible, the data are obtained one at a time, and computational aids are readily available.
- Attributes charts may be used in cases where variables data are not available. In chemical processes this situation frequently occurs in packaging and shipping areas.

33

- Standard variables control charts assume that the population is normally distributed and that samples are independent of each other. Many chemical parameters are, in fact, nonnormal. Samples are sometimes autocorrelated with prior samples. Special treatment is often required.

- Control charts should be used to supplement and complement dynamic process control, not replace it.

- Control charts should also be applied to process and test equipment.

6.5.2 Policy.

To ensure its fitness for use, the product must be made from a consistent supply of material that meets acceptance specifications. The product must be produced by a process that is consistent over time and passed by a statistically valid release-testing procedure. A consistent manufacturing process is required to assure that product characteristics that cannot be tested are consistent from lot to lot. Statistical process control, using the discipline of control charts, is the recommended technique to ensured a consistent process.

6.5.3 Control chart principles.

- The use of control charts represents a commitment to manage the process on "aim" and in a state of statistical control.

- The user has a commitment to investigate for special causes when nonrandom behavior is observed.

- The use of predetermined, statistically based limits determines the appropriate interval for control actions.

- Where appropriate for the process, averaging of a larger subgroup size may be used to increase a chart's sensitivity.

6.5.4 Parameters for control charts.

- Aim (or target). The aim is the goal or aiming point value for the process. It is often called centerline and is usually determined from either the midpoint of the specification range or the long-term mean for the process.

- Standard deviation. The standard deviation of the subgroup size is used for the control chart, not the standard deviation of units of product or the long-term variability. Each control chart procedure includes a method of obtaining this value. Some control charts use the range (multiplied by a constant) within a sample to develop an estimate for standard deviation.

- Control limits. Appropriate action must be initiated when the controlled statistic exceeds the control limit. Each control chart procedure includes a method of obtaining the limits. Multiple decision criteria may be employed. The use of several tests for abnormal patterns is helpful for successful application of the X and X-bar charts.

6.5.5 *The Individuals (X) Control Chart* is a chart of individual values plotted in the form of a Shewhart X-bar chart for subgroup size n = 1. It is a widely used control chart in the chemical industry due to cost of testing, test turnaround time, and the time interval between independent samples. In many cases, the individuals chart is the practical choice because only single observations are available.

6.5.6 *The Moving Range (MR) Chart* shows the absolute difference between the current and previous value, and is often plotted in addition to the X or MA chart.

6.5.7 *The Average (X-bar) Chart* averages several observations together and is more sensitive to small shifts than the X chart. The control limits are narrower since the standard deviation of the sample subgroups are smaller than for individuals.

6.5.8 *The Range (R) Chart* is the absolute difference between the highest and lowest values within a subgroup. The R chart is not appropriate for subgroup sizes larger than 12. The R chart is nearly always plotted together with the X-bar chart.

6.5.9 *The Standard Deviation (s) Chart* is the standard deviation of the results within a subgroup. In the CPI it is appropriate to use the s chart instead of the R chart for subgroup sizes greater than 12.

6.5.10 *The CUSUM Chart:* The CUSUM is the sum of the deviations from target. The CUSUM can be implemented graphically or numerically. In the graphical form a V-mask is used instead of action limits. Whenever the CUSUM trace moves outside the V-mask, control action is indicated. In the numerical form, which is more widely used, the CUSUM is computed from a selected bias away from target value and a fixed action limit is used much as in the Shewhart Chart. Additional tests for abnormal patterns are not required.

6.5.11 *The Moving Average (MA) Control Chart:* The moving average of the last n observations provides a convenient substitute for the X-bar in the Shewhart Chart. However, it does require the user to specify a fixed number of samples, n, for averaging. Choice of the interval may be difficult to optimize. The control signal is a single point out of limits. Additional tests are not appropriate because of the high correlation introduced between successive averages that share several common values. The MA chart is sensitive to trends and is sometimes used in conjunction with X and/or MR charts.

6.5.12 *The Exponentially Weighted Moving (EWMA) Average Chart:* Each new result is averaged with the previous average value using an experimentally determined weighting factor, a. Usually only the averages are plotted and the range is omitted. The action signal is a single point out of limits. The advantage of the EWMA compared to the MA chart is that the average does not jump when an extreme value leaves the moving average. The weighting factor, a, can be determined by either an "effective time period" or by the method covered in Hunter's article (JQT, 1986). Additional tests are not appropriate because of the high correlation between successive averages.

6.5.13 *The Multiple-Variable Control Chart.* Occasionally it is desirable to plot as the control chart variable a calculated parameter constructed from two or more measured variables (e.g., color or tensile tests). The calculated variable may be constructed on the basis of theoretical relationships, principal components, or Hotelling's T^2 (see Kotz, Johnson, and Read in the bibliography). The statistical properties of the combination should be validated before use.

6.5.14 Sampling frequency for control charts.

- *Discrete process operation:* Good practice calls for sampling every batch if batch-to-batch control is desired. Batches may be skipped only after it has been shown that the process is stable over reasonably long periods.

- *Continuous operation:* Frequency of sampling is determined based on knowledge about the process and residence time. This period can be determined from process dynamics or from a statistical analysis of past data. Sampling periods may vary from 15 minutes to 24 hours or more. Intervals should be far enough apart so that samples are not autocorrelated.

6.5.15 Corrective and preventive action.

The occurrence of an action signal indicates that the pattern of the data is not consistent with the level and variability expected. Action should be taken to investigate and/or correct the process promptly. A list of appropriate actions should be available for immediate and uniform implementation. Emphasis should be placed on preventive action not merely compensating action.

6.5.16 Resampling/retesting.

Resampling or retesting not included in the design of the control chart should not be done. Resampling or retesting delays corrective action and changes the risk levels on which the chart control limits have been calculated. The perils of resampling are discussed in chapter 18 of *Specifications for the Chemical and Process Industries.*

6.5.17 Control chart implementation.

Standard Control Chart Applications

Control Variable Variable	Threshold or Limit	Deviation or On-Aim
	— Type of Control —	
Process average	X, X-bar, MA	CUSUM, EWMA
Variability	R, s, MR	CUSUM, EWMA
Percent defective	p	
Number defective	np	
Number of defects	c, u	CUSUM, EWMA

See figure on page 38 for control chart selection.

- Selection of variables: The number of variables that can be maintained on control charts is limited by the facilities and the human attention span. It is generally not wise to put every measurement on a control chart. Priority should be given first to important product or process properties, and second to key control parameters. Charted variables should be controllable by the person(s) maintaining the chart.

  Avoid "forced" variables, those that are made to be within a certain range because of a specification requirement. Look for the hidden variable that compensates for it. For example, the temperature of a household thermostat may remain close to 70°F, while the energy consumption varies.

- Manual charting: Control charts were developed prior to the wide availability of computers, and all techniques can be implemented by hand on paper. This is often the best choice for infrequent measurements, for pilot programs, or for short-term applications. Chart forms can be prepared in advance and operating personnel trained to plot the charts.

- Computer software is available for any control chart application. The use of computers can reduce the work, maintain a data base, allow the keeping of many charts, and assist in diagnostic follow-up by use of graphical output on demand, adequate storage, error-correcting procedures, and checking of assumptions. The features of purchased programs should be understood and validated before use.

6.5.18 Control chart maintenance.

Every control chart application should be periodically reviewed for performance and the estimates of the aim, standard deviation (or range), and the control limits modified as required. Charts with hourly sampling should be reviewed monthly and daily charts reviewed quarterly. The base period for the reference statistics should be at least three review periods if meaningful data are available.

Guide for Control Chart Selection

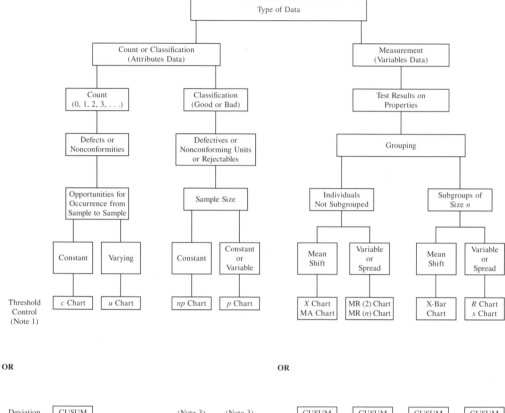

6.6 Process Capability

6.6.1 General.

The process capability concept provides an important means for allocating resources and assessing the effectiveness of programs for quality control and improvement.

6.6.2 Definition.

Process capability is the calculated inherent variability of the product made by a process. It represents the best performance of the process over a period of stable operation.

Process capability is expressed as 6σ, where σ is the standard deviation of the process under a state of statistical control.

6.6.3 Process capability indices.

- C_p

  The capability index, C_p, is the ratio of the specification range divided by the process capability, 6σ:

  $$C_p = \frac{(USL - LSL)}{6\sigma}$$

  where USL is the upper specification limit and LSL is the lower specification limit. This σ is for the short-term component of variation.

  Values of C_p exceeding 1.33 indicate that the process is adequate to conform to the specification limits. Values of C_p between 1.33 and 1.00 indicate that the process, while adequate to meet the specifications, will require close control. Values of C_p below 1.00 indicate that the process is not adequate to conform to the specification limits and that the process and/or the specification limits must be changed.

- C_{pk}

  The capability index adjusted for the process average, C_{pk}, takes into account the possibility of the process average being off center. It is the difference between the process average and the closer specification limit divided by one half of the process capability, 3σ:

  $$C_{pk} = MIN \left[\frac{(USL - PA)}{3\sigma} , \frac{(PA - LSL)}{3\sigma} \right]$$

  where PA is the process average. This σ is for the short-term component of variation.

  As for C_p, high values of C_{pk} are desired. Off-center operation will severely reduce the index and identify the need for an improved control of the process average.

6.6.4 Capability study.

- Evaluation period

  Process capability should be computed from data collected from a period of steady operation in a state of statistical control. If the requirement of statistical control is not precisely met, the process should be operated to a target value using the usual control techniques.

- Data evaluation

  —Control chart:

  If the process is controlled using a conventional X-bar R chart, the process capability can be computed by the formula:

  $$6\sigma \cong 6s = 6\,\frac{\bar{R}}{d_2}$$

  where s   = the calculated standard deviation of the sample

  R-bar = the average range of at least 10 subgroups

  d_2 = the constant for converting ranges into standard deviations for the particular subgroup size used.

  —Frequency distribution:

  If no control chart is used on the property or if some other type of control chart is used and the process is in a state of statistical control, then a sample of at least 30 to 50 observations should be taken and the sample standard deviation, s, computed. The process capability is computed as:

  $$6\sigma \cong 6s$$

  where s = the calculated standard deviation of the sample.

6.6.5 Measurement variability.

Measurement variability is often much larger for properties important in the process industries (e.g., viscosity and purity) than it is for properties measured in the mechanical industries (e.g., dimensions and voltages). Measurement may be responsible for 50 percent or more of the observed variation.

Measurement variability may or may not be included in the estimate of the sigma used for computation of the process capability. It is important that the inclusion or exclusion of the measurement error be clearly intentional and specified in reporting the results.

If the measurement variability is large, care should be taken not to allow the use of sampling plans and sample averages to obscure true process variability. Consistent procedures should be used in computing the process capability for the same property and process at different times in order that the values be comparable.

6.6.6 Process performance.

Process performance is different from process capability. Process performance represents the actual distribution of product variability over a long period of time such as weeks or months, while process capability represents the product variability over a short period such as minutes or hours. Process performance variation will be wider than process capability since it will also include additional components of variation due to time factors.

- Performance study when not in a state of statistical control

  It is sometimes desired to estimate the process performance capability over a long evaluation and/or when not in a state of statistical control (e.g., a customer requires that an index be reported even though control charts show out-of-statistical-control results).

  The performance index, P_p is:

  $$P_p = \frac{(USL - LSL)}{6s}$$

  where s is the calculated standard deviation for the total included population. The performance index adjusted for the process average, P_{pk}, is:

  $$P_{pk} = MIN\ [\ \frac{(USL - PA)}{3s}\ ,\ \frac{(PA - LSL)}{3s}\ ]$$

  P_p/P_{pk} will always be less than or equal to C_p/C_{pk}, respectively.

6.6.7 Variability of Capability and Performance Indexes.

Anyone using capability or performance indexes should be aware of the variability in such indexes. These tables give 95 percent and 99 percent confidence limits for P_{pk} for levels of 1.00, 1.33 and 1.67. These values have been determined by 100,000 simulation each (Kittlitz, 1987). Although computed for P_{pk}, these limits are appropriate for the other indexes as well.

95% Confidence Interval for P_{pk}

n	$P_{pk} = 1.00$		$P_{pk} = 1.33$		$P_{pk} = 1.67$	
30	0.76	1.31	1.02	1.76	1.29	2.19
60	0.83	1.21	1.11	1.61	1.43	2.01
120	0.88	1.14	1.17	1.52	1.47	1.90

99% Confidence Interval for P_{pk}

n	$P_{pk} = 1.00$		$P_{pk} = 1.33$		$P_{pk} = 1.67$	
30	0.71	1.47	0.95	1.95	1.20	2.43
60	0.78	1.30	1.05	1.71	1.32	2.15
120	0.84	1.20	1.13	1.59	1.42	1.99

Note that these confidence intervals are not only wider than many users would like to see, but they are not symmetrical about the nominal value.

6.6.8 Special problems.

- One-sided specifications

  One-sided specifications present special problems. If the practice has been to run as close to the specification as possible, good quality practice requires that an operating target be set at least 3σ from the specification. If a capability index is computed, C_{pk} should be used.

  However, there may be no nominal target value either because goal is as low as possible (e.g., impurities) or as high as possible (e.g., strength).

  In this case, the process capability index concept should not be used to permit intentional contamination or degradation of the product.

- Nonnormal distributions

  The use of the 6σ factor in defining the process capability assumes that the distribution of the property follows the normal distribution. Routine plotting of histograms is a good practice to provide a graphical representation of process data. If it is known or suspected that the normal distribution does not describe a property, the investigator has the following four options

  —Transform.

  Apply a transformation to the data to render the distribution of the transformed property effectively normal. (See Natrella, 1966, chapter 20.)

  —Known distribution.

  Use the multipliers that correspond to the 0.13 percent and the 99.87 percent points for the known nonnormal distribution.

  —Use the index "as is" as a relative measure and so note in report.

  —Omit the computation because the values so obtained would be misleading.

6.7 Design of Experiments

Industrial experiments are expensive to conduct. The proper design of experiments (DOE or DOX) provides the most reliable information efficiently and effectively. The theory behind DOE and the analysis of the DOE can be found in Box, Hunter, and Hunter, 1978.

6.7.1 Screening designs.

Quite often the experimenter is faced with the evaluation of many variables (i.e., 10–25), but limited time to conduct experiments. Screening design experiments are performed first to reduce the number of variables to a more manageable number (i.e., 2–8). Some of the recommended designs are Plackett-Burman (PB) and fractional factorials (FF). These designs will isolate the main variables.

6.7.2 Factorial or fractional factorial designs.

Further understanding may be required to determine if there are interactions involving the main variables or factors. A second set of experiments are conducted using factorial or possibly FF designs. Example: Four variables and two levels require $2^4 = 16$ experiments. From the analysis, the four main effects, the six two-factor interactions, the four three-factor interactions, and the four-factor interaction can be determined.

6.7.3 Response surface models.

If an understanding of possible curvature is required, a third set of experiments is needed, generally known as response surface models (RSM). These experiments are conducted with at least three levels. DOE reference books should be consulted for further guidance.

6.7.4 Mixture designs.

Many chemical products are mixtures of several components or ingredients. Examples include plastics, elastomers, alloys, and lubricants. The desired characteristics depend on the relative proportions of the components in the mixture—not on the total amount of the mixture. Since the percentages of components must add up to 100 percent, the components cannot be varied independently. Special classes of designs are required because of this restriction. Mixture designs are usually set up and analyzed with computer assistance.

6.7.5 Robust designs.

Robust design methods aim at making a process less sensitive to variation in the input factors (raw materials, process parameters, and environmental conditions). This is accomplished by determining settings for the control factors that reduce variation in the response cause by variation in the controllable and environmental factors. These methods include the use of response surface methods and the propagation of error models.

7. New Products/Processes

Summary:

Guidelines for good quality practices should be provided in the area of new products and new processes.

7.1   Sampling of Products Under Development

7.1.1   The development of a new product will typically involve providing samples to potential customers and their testing of candidates prepared only in small pilot or preparative quantities. Full-scale process development and manufacture should begin only after such testing has created a market for the candidate product.

7.1.2   Important characteristics of the material that will be made from the scaled-up process are often unknown at the candidate stage. This situation makes close cooperation between supplier and potential customers especially vital, and they should communicate their respective product requirements (as regards both quality and quantity) and process capability on an ongoing basis during the development process.

7.2   Design Documentation/Process Transmittals

When the decision to commercialize a new product has been made, information necessary for detail design, start-up, and later operation of production facilities should be recorded in a process transmittal. This documentation may be considered confidential. Among items to be included in a process transmittal are:

- Product synthesis chemistry, thermodynamics, and kinetics.

- Projected yields, time cycles, and process conditions at each step of the process.

- Necessary information for process control, including adjustments to be used to meet key in-process and final product properties.

- Physical and chemical properties of products, raw materials, and intermediates, including those related to toxicity, hazards, storage stability, and required materials of construction.

- Tentative targets and limits for products and raw materials, together with test methods for these characteristics. It is recognized that specifications on a newly developed product may be based on limited data and require subsequent revision in the light of better determinations of customer requirements and of process capability.

7.3   Process Improvement

7.3.1   Improvements in the process are required because

- Competition is getting better.
- Costs must be reduced.
- Customers require a better product.
- New government regulations.
- New markets are available if a better product can be made.

7.3.2   Tactics to accomplish process improvement:

- Assessment
  —Benchmarking
  —Process audit
  —SPC for assessment
- Process control and correction
  —SPC to remove special causes and variability
  —Cross-functional teams
  —Producing to target
  —Management review
  —Corrective and preventive action
- Process Improvement
  —DOE
  —Quality function deployment (QFD)
  —Analysis of variance (ANOVA)
  —Benchmarking

8. Shipping

Summary:

- Guidelines should be provided for good quality practices in the area of label control for chemical products. This section does not specifically address the legal, safety, health, environmental, or regulatory aspects of labeling or label control. Container labeling control is aimed at ensuring that each container is correctly labeled as to the identity of the material it contains, the quality of that material, the quantity, and safety hazard information required for the safe handling of the material.

- Good quality practice should be documented concerning the determination and reporting of shelf life as it relates to materials.

- Practices should ensure that product quality is maintained during loading, shipping, and unloading of material products from the producers to the customers.

8.1   Labeling Control

8.1.1   A container is any apparatus—including tank cars, hopper cars, trucks, bags, drums, bottles, boxes, pipelines, storage tanks, holding tanks—in which or through which a material is transported or stored.

8.1.2   Labeling control applies to individual containers that are both the parent lot or any part of the parent lot. Accurate labeling is important at all points in the production, packaging, handling, transport, delivery, and use of materials.

8.1.3   Packaged material.

Where possible, the exact number of containers needed to package the lot should be predetermined and segregated. If physical labels are to be applied to each container then the exact number of labels should be printed and the unique lot identifier included on each label. Otherwise, the label information, including the lot identifier, needs to be marked directly on each container.

8.1.4   Containers that cannot be physically marked should be uniquely identified on records such as bills of lading or the certificate of analysis and timeline control.

8.1.5   Pipelines of materials present labeling problems of their own. In those areas where the pipe cannot be relabeled in a timely or economical manner, labeling should indicate the category of materials carried by the pipe and the hazard warnings, that is, flammable gases, flammable petroleum distillates, crude oil, process water cold–hot, and so forth. At the point of charge and discharge the specific material being transported by the pipe at that time should be clearly displayed. Labels at the charge and discharge points should include the purity, safety, and health warnings for the material.

8.1.6 If a filling operation is interrupted and the product is still conforming, a new lot number should be assigned before packaging is continued if serial marking is not being used, or the test results differ from earlier results.

8.1.7 Continuous production.

For continuously produced products, the most effective system is to date-stamp the final containers because a lot exists only as material produced during a specified period of time. The exact number of labels is defined as a continuous supply for the duration of the packaging operation, with each day's use equaling the number of containers filled.

8.1.8 Inventory records should include individual container serial numbers to allow segregation of portions of a lot.

8.1.9 There should be no shortage of labels that would leave a container without identification for any period of time. There should be no excess of labels that may be used on another lot or another material. All excess labels should be destroyed.

8.1.10 Consumer and industrial products.

Industrial materials should be handled by trained personnel at both the manufacturer's plant and the customer's plant. Labeling of these products is usually concise and technical. Consumer products at their final point of use will be handled and used by people with an unknown degree of understanding of the technical terminology and of materials. Consumer product labeling must be done with considerable care to ensure that the important aspects of a product are clear to the customer.

8.1.11 Contents of a label.

Essential contents

• Name of material
• Safety and health warnings

48

- Net weight
- Lot number
- Manufacturer's name and address
- Date of manufacture
- Shelf life

Nonessential, but useful, contents

- Storage requirements
- Contact number in the event of quality problems or safety issues
- Product code

8.2 Shelf Life Control

8.2.1 Definition.

Shelf life is that period of time during which a material may be stored in the original container under specified conditions of temperature, relative humidity, and/or other physical conditions, and retain expected performance.

8.2.2 Determination of shelf life.

The determination of shelf life is an integral part of product development. For material produced infrequently, adequate data are frequently not available to make a statistically sound determination of the shelf life. One should take into account the real-world storage condition for shelf life determination for commercial products.

8.2.3 Extension of shelf life.

- Since material may still be usable after the specified period of time has elapsed, the testing required to predict shelf life extension should be established by the manufacturer.
- The current owner of the material has the responsibility for the collection of any supporting data for the decision of suitability of the material for use, and for the actual extension.

8.2.4 Reporting shelf life.

The shelf life, where known to be finite, should be part of the labeling of the material. It can be reported via an expiration date or a date after which testing is required to reconfirm the quality of the material.

8.3 Product Shipment

8.3.1 Quality assurance of containers.

- The specific containers to be used should be carefully selected, with consideration given to corrosion, contamination potential, prior contents, in-transit conditions, loading and unloading facilities, product flow characteristics, and labeling needs.

- Procedures should be defined to ensure the integrity of the container (to prevent possible loss or injury) and the cleanliness of the container. Inspection plans and responsibilities of both the container suppliers and the chemical product manufacturer should be defined and communicated.

8.3.2 Product loading for transport.

- New or reconditioned containers should be inspected prior to product loading to ensure the integrity of the container and the absence of manufacturing residues, contaminants, or other foreign material.

- Handling procedures should assure that the potential for contamination is prevented or at least minimized. Sampling of dedicated containers may be required to determine if the heel will contaminate the incoming product. No product should be introduced into a container known or suspected to be dirty or that contains the heel of another product.

- Product packaged in drums, bags, or boxes to be shipped by truck or rail should have appropriate dunnage to prevent damage.

8.3.3 In-transit considerations.

- In-transit product transfers should be avoided. Contamination, damage, and loss are potential risks in these operations. If transfers (particularly from one container to another) are necessary, then the responsible carrier or terminal must be given clear and complete instructions for making the transfer.

- If product (or the container) is damaged or exposed to potential contamination, the owner of the product should be notified immediately. Under no circumstances should spilled or damaged product be repackaged and delivered to the customer without proper authority.

8.3.4 Product unloading.

- The receiver of the product should ensure that container storage, unloading lines, and storage equipment will not jeopardize product integrity.

- If the bulk unloading operation is significantly interrupted on either a planned or unplanned basis, then samples may need to taken and tested to determine if the material is still in conformance with the specifications. The need for a resample and retest is based on the characteristics of the specific material and the specific bulk unloading process. Criteria should be established before the start of the bulk unloading operation. All interruptions need to be documented in the production records for later use in process control and problem resolution.

9. Distributor Relations

Summary:

Good quality practice should be documented concerning materials sold through or purchased from distributors.

9.1 Introduction

Good quality practices are equally applicable to the manufacturer, the customer, the transporter, and the distributor. The distinction between the quality programs needed by each lies in the product-handling functions (packaging, manufacturing, storage, filling, labeling, and bulk transfers) they perform and the responsibilities they have relevant to the quality of the material. The quality programs needed by these functions and the responsibilities are all well documented in ANSI/ASQC 9004-1-1994 and this publication. Each organization must select the subgroup of these quality programs that will fulfill its responsibilities for quality.

The needs of distributors warrant some discussion because their place in the supply chain has had little exposure or discussion.

9.2 General Distributor Quality Program Requirements

Appropriate services and expertise should be provided to suppliers and customers. A distributor should have as a minimum the following verified systems:

- Supplier certification
- Shipping control
- Storage conditions including shelf life management
- Specifications management
- Certificate of Analysis (COA)/Certificate of Conformance
- Material safety data sheet (MSDS)
- Compliance to transportation regulations (for example, DOT)
- Complaints
- Responsible disposal

9.3 Distributor of Packaged Material

Certain distributors serve as the marketing and sales function for the manufacturer. When the distributor handles the materials, they are in the sealed containers as received from the manufacturer.

- The manufacturer has responsibility for product quality until the material reaches the customer and should

—Train the distributor in proper storage and handling of the products.

—Maintain a regular audit and training program to assure that the distributor is capable of representing the manufacturer.

- The distributor needs the general distributor quality program listed in Section 9.2 to determine which manufacturers will provide the necessary quality support. The distributor must also implement those programs the manufacturer has determined are needed to maintain the quality of the material while in the distributor's possession.

9.4 Repackager

Distributors who repackage product receive the material in large quantities and repackage, with the manufacturer's approval and guidance, into smaller packages. These distributors sell the product under the manufacturer's label. Here, the manufacturer again has responsibility for product quality. This distributor needs to add to his quality program a system for documentation of procedures for handling the products during transfer and subdivision, and a system for process control. Safety and disposal concerns also increase in this situation.

9.5 Manufacturer–Distributor

The manufacturer–distributor provides a combination of manufacturing and distribution. Products are received in bulk quantities, repackaged, and sold under the distributor's label. In addition, the products may be mixed further with other products to produce a new product. Also, the same product from different manufacturers may be stored and sold from a common tank. This distributor is responsible for product quality and needs a quality assurance program—meaning a full quality management program of the type described by this manual.

9.6 Warehouser

This distributor serves as the warehouse and shipping arm of the manufacturer. The manufacturer arranges the sales. The quality responsibilities reside primarily with the manufacturer and are listed in Section 9.3.

10. Audit Practice

Good quality practice should be documented concerning the internal or external audit of quality-related issues within an organization.

10.1 Introduction

Quality audits are a major task of the quality function. They are an important tool in the establishment, measurement, maintenance, and improvement of product and systems quality. Quality audits determine the current status and deficiencies of decision making and over time show determination of progress and needs. Quality audits are equally applicable internally and externally. An audit is the objective quantized measurement of an area's capability to do the job assignment. Auditing is applicable to all parts of the supply chain—supplier, manufacturer, transporter, distributor, and customer.

10.2 Types of Audits

There are many types of nonfinancial audits involving quality in the chemical and process industries.

- The most common is a quality system audit that is based on ISO 9000 standards.

- Product-specific audits are another type of audit, often conducted by customers on suppliers. Rabbinical audits to certify food materials and additives as Kosher are a type of product audit.

- Many other types of audits are based on various regulations and regulatory agencies—good manufacturing practices (GMP) audits based on FDA requirements, Department of Transportation (DOT) audits based on transportation regulations, and the like.

10.3 Levels of Audits

Audits may be conducted by the organization on itself (internal audit) or by an outside party (external audit). If the external audit is by a customer or on a supplier, or by a regulatory agency, it is termed a second-party audit. If the external audit is by a neutral party, it is a third-party audit. Certification of a quality system to an ISO 9000 standard requires a third-party audit.

10.4 Auditors

All auditors need to be properly trained, not only in techniques and the standards to which the audit is conducted, but in objectivity and in ethics. The training will vary with the type of audit and the level of auditing.

Generally, there is an audit team with a lead auditor. The lead auditor not only organizes and manages the team, but gives guidance and negotiates any disputes or misunderstandings during the audit.

The requirements for training and experience for qualification as an auditor need to be documented, depending on the level and type of audit. There are ANSI/ASQC standards for auditing quality systems. In addition, there are bodies in many countries (e.g., the Registrar Accreditation Board in the United States) that specify the requirements for training and experience for qualification and certification as an external auditor for ISO 9000 certification

10.5 Conducting an Audit

The generic details of conducting audits are well documented. The specific program developed depends on the needs of the organization. Checklists are a valuable tool for conducting complete and repeatable audits. The sections of this manual are recommended as minimum points to be checked. Areas to be noted especially for CPI are

- Cleanliness of the facility in all areas from production to laboratories and break rooms
- A quality assurance function that promotes cross-functional knowledge
- Testing laboratory processes operating in a state of statistical control
- A process calibration system to ensure consistent operation and maintenance of process capability and performance
- A system for review, documentation, and control of changes to specifications for incoming materials and products, procedures, test methods, and storage conditions
- A quality improvement program
- A sampling program that applies the most relevant knowledge of chemistry, statistics, and the production process
- Adequate storage conditions
- A training program to assure the knowledge needed to perform a given job
- A traceability system
- Audit program existence, format, and schedule

11. Customer and Supplier Relations

Summary:

- A description of good quality practices should be provided in the selection and approval of purchased materials suppliers.

- Good quality practice should be documented concerning the sharing of product data with the customer. The data may take the form of a listing representing the material in the shipment as well as a chart representing some appropriate production interval.

- When quality problems, such as retrievals, complaints, or returns, occur, both the supplier and customer should focus on defining the problem, identifying the root cause, and planning action to prevent the recurrence. Information concerning both the supplier and customer's process that is needed for the resolution of the problem should be available.

- The practice of quality waivers should be discouraged.

11.1 Supplier Approval

11.1.1 Customer–supplier relationship.

Because raw materials are major components of many processes, they can be a significant source of both process bias and special cause variation. It is highly desirable to consider the raw material supplier as a partner and develop a cooperative relationship. A confrontational relationship benefits neither the customer nor the supplier. While many supplier relationships involve sources outside the user's organization, everything that follows can be and should be applied to suppliers within the organization as well.

11.1.2 Approval process.

- The identification of prospective suppliers includes consideration of the following factors

  —Previous experiences and relationships with the supplier

  —Supplier's area of manufacturing expertise

  —Supplier's quality assurance program

- Once prospective suppliers have been identified, the proposed material specifications with their related test and sampling methods, are provided. This is also an appropriate time to discuss, in detail, the supplier's quality system.

- Once the field has been narrowed to the most qualified suppliers, discussions should occur in order to share more detailed information concerning the purchased material and the process needs of the customer.

- Representative samples can now be requested and evaluated for conformance with specifications and for stability.

- Following laboratory evaluation of the samples, larger quantities of material may be utilized in pilot plant trials, limited plant trials, or extended plant trials.

- At this point, quality audits of the suppliers are appropriate. Any prior audits of the suppliers are reviewed and updated where needed. An on-site audit may be needed.

- During this approval process, the target and limits are refined.

- Based on a supplier's agreement to the specifications, the supplier is approved.

- The complete supply chain needs to be included in the approval process.

11.1.3 Functions and responsibilities.

In most organizations, the responsibility for the process of supplier selection rests with the purchasing function. It is essential that this be a coordinated team effort between purchasing, R&D, process development, process engineering, production, quality assurance, and the supplier. The responsibilities of these team members are listed as follows

- Purchasing acts as coordinator of this team process, which brings together the potential suppliers and the users.

   —Keeps communications current with the suppliers concerning the approval process, specifications, quantities, and schedules.

   —Participates, where agreed, in any audits of the suppliers.

   —Maintains a list of approved suppliers.

- The supplier is responsible for providing information and data describing the product and process and for providing help to identify critical parameters that will ensure satisfactory performance.

- Product design functions (R&D, process development, and process engineering) evaluate candidate raw materials both at the laboratory and pilot plant levels and define acceptable ranges for specification limits.

- Production conducts final evaluations of purchased materials and may perform in-plant production trials.

- Quality management coordinates the preparation of purchase specifications and leads the team that audits the supplier's facilities and quality systems. Quality management assures that all steps have been completed and thus forms a total process that serves as a basis for measurement of the relationship.

11.1.4 Single versus multiple sources.

While multiple sources may provide some assurance of a continuous supply of material, they can also result in greater variability in the final product. In processes where variability must be minimized, the use of a single source may help. While the selection of a single source may provide decreased variability, the selection must be done carefully and should involve mutual commitment by the supplier and organization to the future of the partnership.

11.2 Data Sharing

11.2.1 Preparations for data sharing.

- A discussion should be held by the parties involved before any data sharing is initiated. This discussion should address the following

  —The supplier data to be shared with the customer

  —The customer data to be shared with the supplier

  —The use of the data, including benefits to both parties

  —The availability of data requested by either party

  —Mechanism for data transmission including format and frequency

  —Issues of proprietary information

  —Complete understanding of the nature and limitations of the data to avoid misinterpretation

- Secrecy agreements may be needed if proprietary information is involved. In addition, the need for any special handling of such information should be clearly defined.

11.2.2 Process control data.

- Provides demonstration of effectiveness of process control.
- Most useful when presented in the form of control charts or time-sequenced raw data.
- Knowledge of measurement variability in data is essential.

11.2.3 Product property data.

- Reports of product properties as measured.
- Accurate dating of data is required for product properties in order to interpret changes with time. The time relationship between supplier's process control, product property, shipments, and customer's performance data may be needed to obtain full benefit from data sharing.
- Knowledge of measurement variability in data is essential.

11.2.4 General review of data.

- A review of the data (quarterly, semi-annually, or annually) should take place for both the supplier's and the customer's mutual interest. Such meetings may provide a working relationship to solve mutual problems.

- Correlation between supplier's and customer's data may clarify specifications, problems, and the need for different data.

11.3 Retrieval of Material

11.3.1 Retrieval is the removal of material from the transportation and distribution network, or from customers, initiated by the manufacturer. The decision to retrieve material may result from the manufacturer's concern over the safety or performance of the product, or in response to regulatory action. If a retrieval is a result of regulatory requirements, it is a "recall," and the specific requirements of the applicable regulations must be fulfilled.

11.3.2 Authority for initiating retrieval procedures should be clearly defined within the manufacturer's organization.

11.3.3 The success of a product retrieval depends on availability of complete records concerning the transportation and distribution of material that may be subject to retrieval. A manufacturer should maintain a system that will permit determination of the amount, date, and destination of all material of a given lot or series of lots, or all material made within a specified time frame. A distributor should keep records adequate to enable the customers of material from any lots distributed—together with the amounts and dates of shipment—to be identified.

11.4 Complaints

11.4.1 A complaint report is a link between a customer's problem and those individuals in the supplier organization who can diagnose and correct the problem.

Complaints can be measures of dissatisfaction. However, a low rate of complaints does not necessarily mean customer satisfaction. Study of complaints, while beneficial, is not a substitute for market analysis.

11.4.2 Classification.

Because complaints can and will cover many aspects of a product, they should be classified by type:

- Obvious errors
- Failure to conform to specifications
- Performance problems
- Packaging
- Order problems
- Scheduling
- Delivery problems
- Inadequate specifications

11.4.3 Complaint report.

The complaint report should contain the following elements

- Identity
    - —Identity of the product/shipment involved
    - —Product name
    - —Time and date the complaint was received
    - —Manufacturing location
    - —Shipping location
    - —Shipping date
    - —Lot number
    - —Customer and supplier order number
    - —Customer name
    - —Customer location
    - —Name of author of complaint report, date of receipt of customer complaint
    - —Date of customer notification and any special requirements for response timing
    - Sample (if any) source and to whom sent
    - —Required response date
- Brief statement of product application
- Complete statement of problem
    Relevant data, test results, and test method should be included. If product performance in an application is involved, lot numbers of satisfactorily performing material should be provided, if available.
- Amount/value
    - —Statement of the actual quantity of material involved
    - —Value of claim or possible claim

11.4.5 Response should contain the following elements

- Direction to the individual who can perform or guide diagnosis of problem, determine cause, and recommend corrective action. This person is responsible for coordinating investigation and preparing response. Fast and complete response to a customer is imperative.
- Diagnosis of problem.
- Root cause of problem.
- Corrective and preventive action.

11.4.6 Management review.

Complaints should be subject to periodic management review. The complaints, the responses, and the corrective and preventative action taken should be reviewed to identify patterns that may identify more effective corrective and preventative actions.

11.5 Returned Product

Product may be returned for various reasons—overstock, complaint, shelf life, goodwill, and the like.

- There must be a procedure of the acceptance and disposition of returned product.
- Upon receipt of returned material, the appropriate individuals or department must be notified.
- Material should be inspected for packaging integrity.
- Materials review board (including of both production and nonproduction personnel) should recommend disposition.
- Appropriate disposition should be taken: inventory, rework, discount sale, scrap, and so forth.

11.6 Quality Waiver

It is necessary to clearly document the details of the shipment, including the reason for the waiver, and to inform all affected parties before the product is shipped.

11.6.1 Shipment of product that does not meet specification is NEVER GOOD QUALITY PRACTICE. Any such shipment must be made under waiver.

Any practice concerning a quality waiver should be equally applicable for intercompany shipments, intracompany shipments, in-process intermediates, and product returns, and should be documented.

11.6.2 If a product has been determined not to meet specifications for one or more specification properties, then the product can be shipped only if all the following criteria are met

- None of the nonconforming properties is known to adversely affect the customer with respect to performance, reliability, or safety.
- The customer's end use is not adversely affected by the nonconformance. This question should be answered by the customer whenever possible.
- Agreement to ship the nonconforming product has been obtained from the customer or the marketing group responsible.
- The material review board, if in place, has approved shipment as a proper disposition.

12. Bibliography

- ANSI/ISO/ASQC A3534-1-1993 *Statistics—Vocabulary and Symbols—Probability and General Statistical Terms.* Milwaukee: American Society for Quality.

- ANSI/ISO/ASQC A3534-2-1993 *Statistics—Vocabulary and Symbols—Statistical Quality Control.* Milwaukee: American Society for Quality.

- ANSI/ISO/ASQC A8402-1994 *Quality Management and Quality Assurance—Vocabulary.* Milwaukee: American Society for Quality.

- ANSI/ISO/ASQC Q9000-1-1994 *Quality Management and Quality Assurance Standards—Guidelines for Selection and Use.* Milwaukee: American Society for Quality.

- ANSI/ISO/ASQC Q9000-2-1993 *Quality Management and Quality Assurance Standards—Generic Guidelines for the Application of ANSI/ISO/ASQC 9002, and 9003.* Milwaukee: American Society for Quality.

- ANSI/ISO/ASQC Q9001-1994 *Quality Systems—Model for Quality Assurance in Design, Development, Production, Installation, and Servicing.* Milwaukee: American Society for Quality.

- ANSI/ISO/ASQC Q9002-1994 *Quality Systems—Model for Quality Assurance in Production, Installation, and Servicing.* Milwaukee: American Society for Quality.

- ANSI/ISO/ASQC Q9003-1994 *Quality Systems—Model for Quality Assurance in Final Inspection and Test.* Milwaukee: American Society for Quality.

- ANSI/ISO/ASQC Q9004-1994 *Quality Management and Quality System Elements—Guidelines.* Milwaukee: American Society for Quality.

- ANSI/ISO/ASQC Q9004-2-1991 *Quality Management and Quality System Elements—Guidelines for Services.* Milwaukee: American Society for Quality.

- ANSI/ISO/ASQC Q9004-3-1993 *Quality Management and Quality System Elements—Guidelines for Processed Materials.* Milwaukee: American Society for Quality.

- ANSI/ISO/ASQC Q9004-4-1993 *Quality Management and Quality System Elements—Guidelines for Quality Improvement.* Milwaukee: American Society for Quality.

- ANSI/ISO/ASQC Q10011-1-1994, Q10011-2-1994, and Q10011-3-1994 *Guidelines for Auditing Quality Systems.* Milwaukee: American Society for Quality.

- ANSI/ISO/ASQC Q10013-1995 *Guidelines for Quality Manuals.* Milwaukee: American Society for Quality.

- ANSI/ASQC Z1.4-1993 *Sampling Procedures and Tables for Inspection by Attributes.* Milwaukee: American Society for Quality.

- ANSI/ASQC Z1.9-1993 *Sampling Procedures and Tables for Inspection by Variables for Percent Nonconforming.* Milwaukee: American Society for Quality.

- Arter, Dennis R. *Quality Audits for Improved Performance,* 2nd ed. Milwaukee: Quality Press, 1994.

- ASQC Chemical and Process Industries Division, Chemical Interest Committee. *ISO Guidelines for the Chemical and Process Industries,* 2nd ed. Milwaukee: Quality Press, 1996

- ASQC Chemical and Process Industries Division, Chemical Interest Committee. *Specifications for the Chemical And Process Industries—A Manual for Development and Use.* Milwaukee: Quality Press, 1996

- ASQ Quality Audit Division. *The Quality Audit Handbook.* Milwaukee: Quality Press, 1997.

- Bishara, Rafik H., and Michael L. Wyrick. "A Systematic Approach to Quality Assurance Auditing." *Quality Progress,* December 1994, pp. 67–70.

- Bossert, James L. *Quality Function Deployment: A Practitioner's Approach.* Milwaukee: Quality Press, 1990.

- Box, George E. P., and Norman R. Draper. *Empirical Model-Building and Response Surfaces.* New York: John Wiley & Sons, Inc., 1987.

- Box, George E. P., William G. Hunter, and J. Stuart Hunter. *Statistics for Experimenters.* New York: John Wiley & Sons, Inc., 1978.

- Burr, John T. "Keys to a Successful Internal Audit." *Quality Progress,* April 1997, pp. 75–77.

- Eagle, Alan R. "A Method for Handling Errors in Testing and Measuring." *Industrial Quality Control,* March 1954, pp. 10–15.

- Fiorentino, Raphael, and Michel Perigord. "Going From an Investigative to a Formative Auditor." *Quality Progress,* October 1994, pp. 61–65.

- Gunst, Richard F., and Robert L. Mason. *How To Construct Fractional Factorial Experiments.* Milwaukee: ASQC Quality Press, 1991.

- Grant, E. L., and Richard S. Leavenworth. *Statistical Quality Control,* 5th ed. New York: McGraw-Hill, 1980.

- Hunter, J. Stuart. "The Exponentially Weighted Moving Average." *Journal of Quality Technology* 18 (1986), pp. 203–209.

- ISO A3534-3-1993 *Statistics—Vocabulary and Symbols—Part 3: Design of Experiments.* Milwaukee: American Society for Quality.

- Johnson, N. L., and F. C. Leone. "Cumulative Control Charts." *Industrial Quality Control* (June, July, August 1962).

- Juran, J. M., F. M. Gryna, Jr., and R. S. Bingham, Eds. *Quality Control Handbook,* 3rd ed. New York: McGraw-Hill, 1979.

- Kirk-Othmer. "Sampling," Vol 20 of *Encyclopedia Of Chemical Technology,* 3rd ed. New York: John Wiley & Sons, Inc., 1982, pp. 525–548.

- Kittlitz, R. G. "PPK Distribution." Seaford, DE: DuPont Report, 1987.

- Kotz, Samuel, Normal L. Johnson, and Campbell B. Read, "Hotteling's T^2." Vol. 3 of Encyclopedia of Statistical Sciences. New York: John Wiley & Sons, Inc., 1983, pp. 669–673.

- Lucas, J. M. "The Design and Use of V-Mask Control Scheme." *Journal of Quality Technology* 8 (1976), pp. 1–8.

- Montgomery, Douglas C. *Design And Analysis Of Experiments,* 4th ed. New York: John Wiley & Sons, Inc., 1997.

- Montgomery, Douglas C. *Introduction To Statistical Quality Control,* 3rd ed. New York: John Wiley & Sons, Inc., 1996.

- Natrella, Mary G. *Experimental Statistics—Handbook 91.* Gaithersburg MD., National Institute of Standards and Technology, 1966.

- Nelson, L. S. "Interpreting Shewhart X-Bar Control Charts." *Journal of Quality Technology* 17 (1984), pp. 114–116.

- Registrar Accreditation Board. *Certification Program for Auditors of Quality Systems (RAB Handbook C1.4).* Milwaukee: Quality Press, 1996.

- Rice, Craig M. "How to Conduct an Internal Quality Audit and Still Have Friends." *Quality Progress,* June 1994, pp. 39–41.

- Russell, J. P. and Terry Regel. *After the Quality Audit: Closing the Loop on the Audit Process.* Milwaukee: Quality Press, 1996.

- Ryan, Thomas P. *Statistical Methods for Quality Improvement.* New York: John Wiley & Sons, Inc., 1989.

- Sahrmann, Herman. "1986/87 Annual Trend Forecast." *Quality Progress,* April 1987, p. 38.

- Schilling, E. G. *Acceptance Sampling in Quality Control.* New York: Marcel Dekker, 1982.

- Shewhart, W. A. *Economic Control of the Quality of Manufactured Product.* New York: D. Van Nostrand Company, 1931. (Republished by AQC, 1981.)

- Tsuda, Yoshikazu and Myron Tribus. "Planning the Quality Visit." *Quality Progress,* April 1991, pp. 30–34.

- Tukey, John W. *Exploratory Data Analysis.* Reading, MA: Addison-Wesley, 1977.
- Velleman, Paul F., and David C. Hoaglin. *Applications, Basics, and Computing of Exploratory Data Analysis.* Boston: Duxbury Press, 1981.

Index